P9-DTA-737

Karen Cook
Staff

BREAKTHROUGHS IN SCIENCE

INVENTIONS

BREAKTHROUGHS IN SCIENCE

INVENTIONS

CAROL J. AMATO

ILLUSTRATIONS BY STEVEN MOROS

Dedication

To space station Freedom and the future of humanity as a space-faring civilization.

Acknowledgments

I would like to thank Rockwell International, Downey, California, for the information on the space shuttle and space station, and Hughes Aircraft Company, El Segundo, California, for the information on weather and communications satellites.

A FRIEDMAN GROUP BOOK

This edition published in 1992
by SMITHMARK Publishers Inc.
112 Madison Avenue
New York, New York 10016

ISBN 0-8317-1013-6

BREAKTHROUGHS IN SCIENCE: INVENTIONS
was prepared and produced by
Michael Friedman Publishing Group, Inc.
15 West 26th Street
New York, NY 10010

Editor: Dana Rosen
Art Director: Jeff Batzli
Designer: Lynne Yeamans
Photography Researcher: Daniella Jo Nilva
Illustrator: Steven Moros

Typeset by Bookworks Plus
Color separations by Rainbow Graphic Arts Co.
Printed and bound in Hong Kong by Leefung-Asco Printers Ltd.

SMITHMARK Books are available for bulk purchase for sales promotions and premium use.
For details write or telephone the Manager of Special Sales, SMITHMARK Publishers Inc.,
112 Madison Avenue, New York, New York 10016. (212) 532-6600.

Table of Contents

INTRODUCTION

❑ ❑ ❑ ❑ ❑

Throughout time, humans have used their superior intelligence to invent. Invention, or the creative spirit, is a uniquely human trait, setting humankind apart from other species. Beginning in ancient times, men and women sought new materials and techniques to help them survive in a hostile environment. As the process of invention continued, civilizations flourished. Many of the first inventions, such as tools and fire, helped lay the groundwork for future inventions. Later, inventions like the wheel and ships helped people travel to new places. Writing allowed men and women to record information, and the printing press relayed their words to a wider audience. With the invention of agriculture, people could settle in one place to grow plants instead of having to roam the earth in the search for food.

Today, we have some of the most exciting inventions of all time—rockets, the space shuttle, and the space station. These inventions have taken humans into outer space and have sent us to the moon and back. With the space station, humans will be able to *live* in outer space.

How did we get from the inventions of our primitive ancestors to the invention of a space station? Each invention is a building block for the next. This book discusses major breakthroughs in four areas—technology, transportation, communication, and living modes—and shows how these wonderful inventions have changed our lives.

Courtesy NASA

This drawing shows how the space station *Freedom* will look when it is complete. The large rectangles sticking out from the long beam are solar panels, which will supply *Freedom* with electric power.

Breakthroughs in Technology

Technological advances have brought humans from the Stone Age to the Space Age. Some of the major technological breakthroughs range from the invention of tools and the discovery of fire, to the harnessing of steam and electric power, to the use of lasers in medicine and robots in factories and homes.

An invention begins when we ask questions about a particular problem. First, we try to answer the questions, often simply by guessing. Then, we think of various solutions to the problem. When we have the best solution, we produce the invention. The innovation that results often changes the way in which we live.

TOOLS

The first humans roamed the earth, living in various grassy areas in Africa called *savannas.* But food was scarce, and so in order to eat, humans learned to hunt animals. The development of any invention depends on need. In order to hunt successfully and to skin the animals they killed so they could eat the meat, humans developed the skill of making various kinds of tools. We know that early humans used tools because many flint objects and tools made of animal bone have been discovered with human remains dating back to prehistoric times.

© Peabody Museum, Harvard University/photo by Hillel Burger

This flint hand axe was made by Neanderthal man, the immediate forerunner of modern man. Found in France, it dates from the Pleistocene epoch, about 100,000 years ago.

The first tool-users were a group of extinct humans we now call the Australopithecines. These apelike ancestors of *Homo sapiens* were about four feet (1.2 m) tall and lived in Africa. Prior to the Australopithecines, our ancestors ate leaves and lived in trees. As the climate changed and the temperature grew warmer, the trees began to disappear, and these ancestors were forced to come down from the trees to search for food. The Australopithecines learned to dig up roots and kill small game. Their first tools were

sticks and stones thrown at animals. Humans later adapted these sticks and stones into clubs and rocks that allowed their owners to hunt farther away from the trees.

Using a stone or a stick as a tool only made humans tool-users, not tool-makers. But, about two million to 250,000 years ago, humans began changing the shape of those stones or sticks into rough tools used to chop meat and crack bones. Many of these chipped stones look like axes. As time went on, they turned the stones and sticks into knives and scrapers and arrowheads and bowls. These tools were the first inventions, and they turned humans from tool-users into tool-makers.

The best stone for tools was flint, because it split easily into flakes when struck. The most useful tool was the hand ax, because it could be used as a hammer, trowel, knife, or scraper. Flint-tipped arrows and spears made humans excellent hunters. They hunted in groups to kill large animals like reindeer, bison, and woolly mammoths.

The invention of tools had an enormous impact on civilization. Many of these first tools became the basis for future, more complex inventions.

Opposite page: The Australopithecines were the first tool-users. These early humans used rocks to crush nuts and bones, as well as sticks to knock food and game from the trees.

FIRE

Humans learned how to use fire over 750,000 years ago. The ancestor who made this discovery is called *Homo erectus*. Bones of *H. erectus* have been found in a cave near Choukoutien (Choo-koo-tee-en), China. The site is near Peking (now called Beijing), so *H. erectus* is often called "Peking Man." Traces of *H. erectus* have also been found in Java and in the Olduvai Gorge in Africa. Archaeologists know that Peking Man used fire, because they discovered charcoal fragments and burned bones in the same layers of dirt in which they found the bones of Peking Man.

H. erectus began living in the warmer tropical zones, but these humans eventually spread into China, where the weather was very cold. They needed fire to keep warm, to cook the meat they ate, and for light. No doubt humans were afraid of fire at first.

Although we really do not know how humans first learned to use fire, scientists have many theories. Some archaeologists believe that these early humans borrowed burning twigs from brush fires caused by lightning. Once started, the fire was kept going or relit from another fire.

But many archaeologists agree that humans learned to create fire by accident. Maybe a spark landed in some dry leaves and began to smolder while someone was chipping flint tools. Or, perhaps they discovered fire by rubbing two sticks together.

Cooking was probably invented by accident, too. Maybe some meat fell into the fire and tasted better after it had been charred.

Homo erectus learned to use flint to chip rocks and create sparks for a fire. Without fire, this early species of human could never have survived the cold weather in the areas of China where they lived.

Fires were probably used in hunting, also. Archaeologists have found burned soil behind areas where kills were made, leading them to believe that fires were set in order to frighten animals into bogs or pits, over cliffs, or toward a band of hunters with spears.

One thing is certain, however: Without the knowledge of the use of fire during the Ice Ages, 300,000 years ago in Europe, humans would not have survived.

THE STEAM ENGINE

When we think of the steam engine, we usually think of it as having been invented in recent history. The first steam engine, however, was invented in ancient Greece by a man named Hero about 120 B.C.

Hero called his machine an *aeolipile* (ay-oh-lih-pile). This name is from Aeolus (Ay-oh-lus), the Greek god of the winds. A covered caldron made steam, which passed through a hollow ball and then escaped through thin, bent pipes, causing the ball to spin. But no one, not even Hero, realized that his machine could be used for power.

Over 1,700 years passed before a man in France named Denis Papin (Day-*nee* Pah-*panh*) invented another version of the steam engine. His had a cylinder and piston, just as the ones used for industry had seventy-five years later. Although Papin got the piston to move up from the pressure of the steam he forced into the cylinder, he, too, had no idea how his invention could be used.

The first person to use a steam engine for practical purposes was an English blacksmith from Devonshire named Thomas Newcomen (1663–1729). In 1712, borrowing Papin's ideas, Newcomen designed an engine to pump water from the sea. Newcomen's engine was also used to pump water out of coal mines, so the miners could work inside.

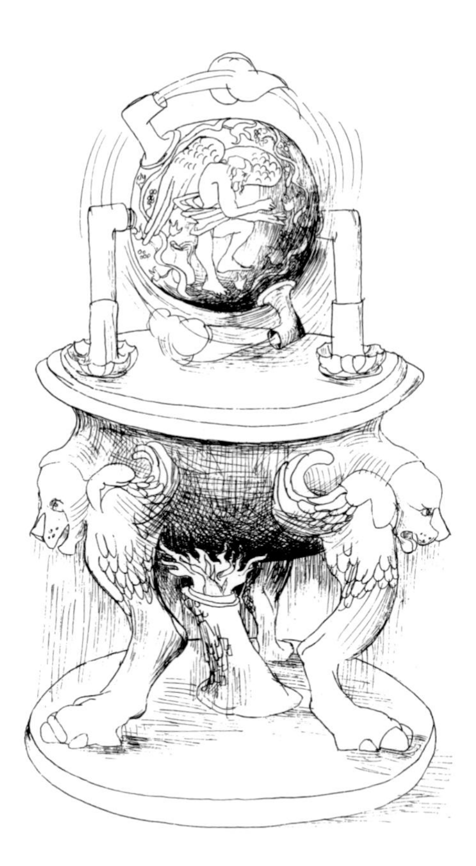

This picture shows Hero's aeolipile, the first engine powered by steam. The bottom container was partly filled with water and then heated. This heat produced steam, which went up the pipes on the side and into the metal sphere on the top. The metal sphere had exhaust pipes with openings at right angles. When steam came out of these openings, the sphere rotated.

Opposite page: Thomas Edison tested 6,000 types of filaments in his light bulbs before discovering that bamboo fiber worked the best.

Newcomen's design, however, had some problems. The engine used a great deal of coal and was expensive to run. It could not be cooled fast enough, and the piston only went up and down twelve to fifteen times per minute.

One day, a Scottish engineer named James Watt (1736–1819) was asked to repair one of the Newcomen engines. Watt, the son of a merchant from Greenock, Scotland, was trained as a mathematical instrument maker and machinist. He soon realized that there were problems with Newcomen's engine, and he tried to improve the design. Several years passed before he came up with the right answer. Watt decided that steam should not only force the piston up, but should force it down again, too.

By 1765, James Watt's steam engine was ready for use. Watt eventually formed a partnership with Matthew Boulton, who was from Birmingham, England. Watt and Boulton became famous for the improved steam engine.

Before the steam engine, horses were used to drive mill machinery. Watt introduced the term "horsepower" to measure the rate at which his engines worked. He arrived at that rate by calculating the performance of the average horse and comparing it to the output of his engines.

The steam engine was the power source that made industrialization possible. Steam engines not only powered factories, but later locomotives and ships as well.

THE ELECTRIC LIGHT BULB

Whenever we need light, we take for granted our ability to walk to the wall switch and flip it on. But just 100 years ago, there were no electric lights.

During the days when people lived in caves, fire was the only source of light besides the sun. This remained true until several hundred years ago when the oil lamp replaced

The modern filament light bulb is much smaller than the one Thomas Edison invented. It does not have a bamboo filament, however, but one made of tungsten wire.

blazing torches or candles. In the late 1800s, gas lighting replaced oil lamps, but it soon gave way to an extraordinary invention still in use today: electricity.

History credits Thomas Alva Edison (1847–1931) with inventing the electric light bulb, but in reality, Joseph Swan was the first to invent an electric filament lamp. At the same time, in 1878, Edison applied for a patent, which is a grant by the government assuring an inventor the sole right to produce an invention. He also formed the first central electric power plant in the world, called the Edison Electric Light Company.

To invent an electric filament lamp, Edison studied work done by three Englishmen. The first was Michael Farraday, who found a way to produce electricity in 1831. Edison also built upon the achievements of Sir Humphrey Davy, who invented a carbon arc lamp, and Sir William Groves, who invented an electric light with a platinum filament. None of these lamps burned for more than a few hours, however, because the filaments—the thin wires inside the lamp—were not made of the right material.

Edison experimented with 6,000 different types of filaments before he made the discovery that bamboo fiber worked the best. He was granted a patent on January 27, 1880, and he demonstrated his electric bulb in Menlo Park, New Jersey, that same year. In 1883, Edison and Joseph Swan formed the Edison and Swan Electric Company, which was located in London, England.

Edison knew people would not use electricity unless it was better and cheaper than gas. He designed a central power station because his bulbs required a constant flow of electricity to keep burning. By the early 1900s, the industrialized world was rapidly converting to electricity as its main form of power. New York was the first city to be powered by electricity.

THE LASER

The word "laser" is an *acronym* (a word formed by the first letters of other words) for *L*ight *A*mplification by *S*timulated *E*mission of *R*adar.

American scientist Charles Hard Townes, and Arthur Schawlaw, his brother-in-law, first suggested the laser theory in 1958. In 1960, American physicist Dr. Theodore Maiman developed the first laser. Ever since then, the use of lasers has grown rapidly.

Laser light, the most powerful source of light we have, differs from electric light in four ways. First, laser light has "directionality"—its beam appears to be the same width all over. Second, lightwaves from a specific laser are all the same length and speed. This is called "coherence." Third, laser light is "intense"—it sends a lot of power to a small area. Finally, laser light is also "monochromatic," which means it has only one color.

Laser beams can be so narrow they form knives sharp enough to go through the skin without causing bleeding or leaving a mark. They are germ-free and can destroy damaged tissue while leaving the surrounding healthy tissue alone, making them very useful for surgery. With lasers, operations can take seconds instead of hours.

Lasers can weld together parts that nothing else can. They can also burn a hole through a diamond or sheet of steel; carry 800 million telephone conversations; measure distances accurately; destroy missiles; patch holes better than thread; and reach temperatures of 11,000 degrees F (6,000 °C)—hotter than the sun's surface.

Many scientists hope that lasers, with their powerful temperatures, will someday provide us with unlimited energy via nuclear fusion, a much safer source of energy than today's nuclear power.

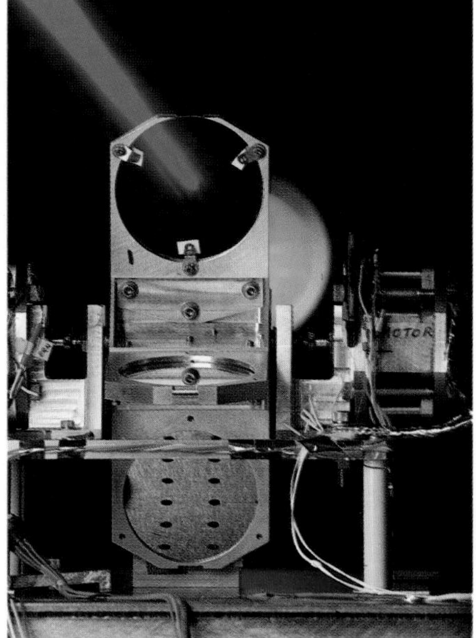

Top: This laser machine is emitting a beam of light. Bottom: These laser beams are cutting a lens. Unlike light from a lamp or a flashlight, which "diffuses" outward in a triangular shape, the laser beam travels in a straight line.

These robot hands are working on a computer keyboard. They have as much wrist and joint action as human hands. Robot hands are also used in factories to handle dangerous chemicals and other products that are harmful to humans.

ROBOTS

A robot is a machine that can do work without a human having to run it all the time. The term "robot" was invented by Czechoslovakian writer Karel Capek in his play, *R.U.R.* (Rossum's Universal Robots). Robot comes from the Czech word for "work." Capek used the term to describe a mechanical being that looks human, but has no emotion and only performs automatic operations.

When we think of robots, we often think of a being like C3PO from *Star Wars.* The idea of metal figures coming to life has been with us since the days of the Greeks! Over the years, people have made mechanical figure toys. In 1932, a robot named Alpha, who could read, bow, tell time, and smoke cigars, was on display in London. At the 1939 New York World's Fair, a robot named Elektro could obey spoken words, walk, count on his fingers, and give orders to his robot dog, Sparky. Sparky was able to wag his tail and bark.

While we are well on the way to creating something like C3P0 at some point in the future, we are at the same stage of robot development that the Wright Brothers were with their World War I airplanes! Although they can perform various tasks, robots today are quite nonhuman.

The origins of today's robots began in the 1950s, and improvements on the first robots were made possible by industrialization and the invention of computers. Robots designed to do repetitive tasks that humans find boring, like assembly line work, or dangerous, like picking up hot metal, have had an enormous impact on modern society. Robots have even travelled to Mars! The robotic *Viking* landers picked up soil and performed scientific experiments on Mars. The space shuttle has a robot arm, used for launching and retrieving satellites.

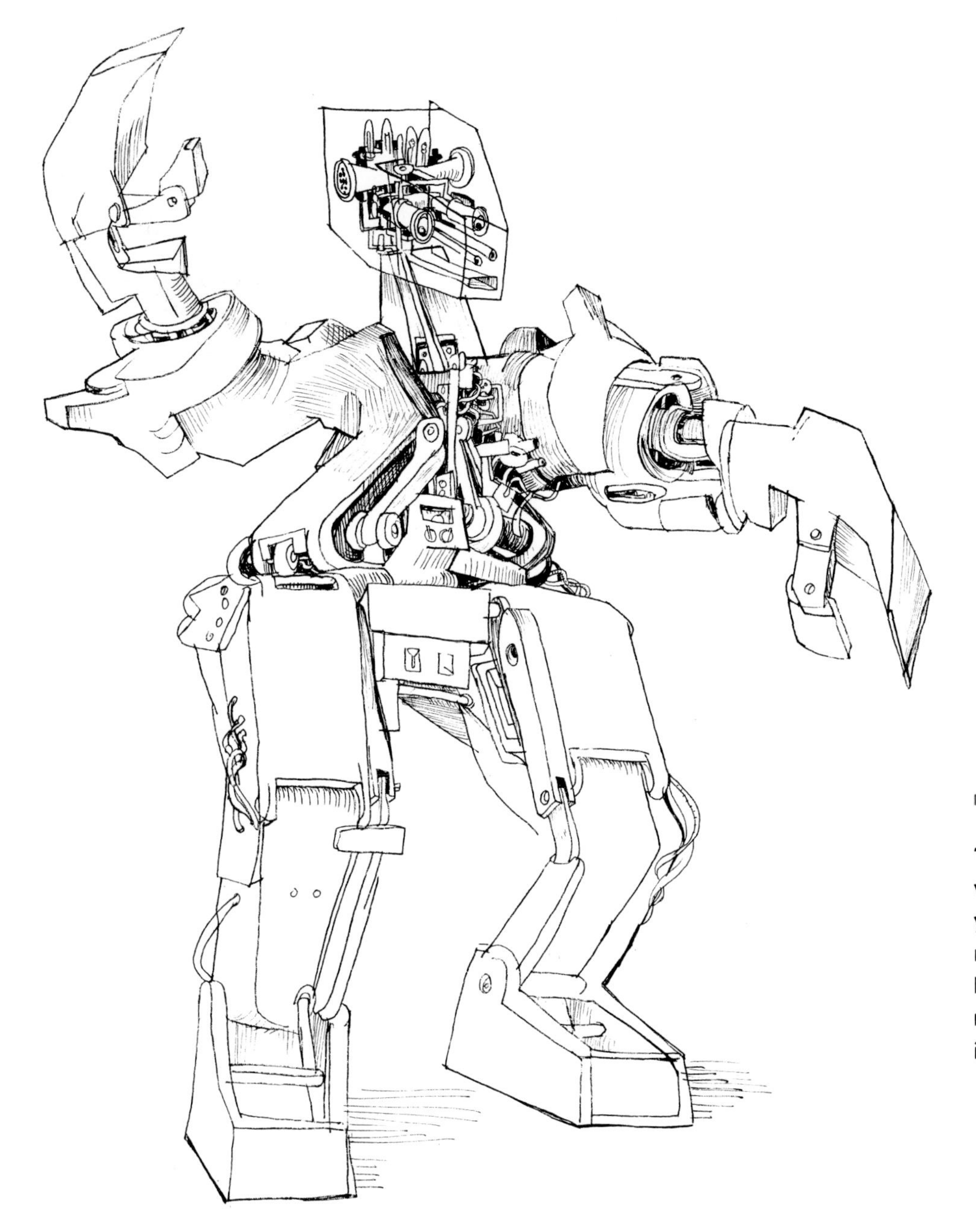

This picture shows a robot designed with a human body form. For many years, people have thought of mechanical beings as being shaped like us. Making them walk with a natural gait and without falling over is a real challenge.

In industry, robots can move around, "see," and perform tasks, even though they do not look human. They are on wheels or tracks, and have television cameras for eyes and computers for brains. The advantage is that robots never get tired, bored, or hungry, and after the initial cost of buying a robot, they work for free.

Do you know that *you* use robots in your everyday life? Your washing machine, dryer, dishwasher, and clock radio are all forms of robots. Perhaps someday we will have robots that truly look, as well as act, human.

Breakthroughs in Transportation

Before the invention of the wheel, people had to travel on foot and carry things by hand or drag them on sleds.

The wheel changed our lives and revolutionized the course of civilization. As one invention builds on another, so the wheel led to other inventions in transportation, like carts, steamboats, trains, and cars.

This carving from early Mesopotamia shows people using wheeled carts. The first wheels were solid circular disks. Spoked wheels came along about 1,000 to 1,500 years later.

THE WHEEL

The earliest known wheels were constructed in ancient Mesopotamia and date from about 3500 to 3000 B.C. Archaeologists believe that they were invented in several places at the same time.

The potter's wheel was invented first, which allowed people to make clay pots. Archaeologists think that people then realized the wheel could be put on their sleds. They invented wheeled carts, which could carry much more than unwheeled sleds and soon became very common.

The first wheels were solid wooden disks made from tree trunks or three planks of wood clamped together. Wheels were mounted on a round axle, and fastened by a wooden pin through the axle. These wheels were very heavy, but not until 2000 B.C.—1,000 to 1,500 years later—did people learn to cut sections out of the wheel to reduce its weight.

The invention of the wheel was a major turning point in the advance of human culture. Because of the wheel, humans could use animals like oxen and horses for farming and other heavy work. The wheel led to increased trade between villages and towns.

SHIPS

Preindustrial cultures first used rafts, and later, canoes to travel by water. As far back as 3000 B.C., the ancient Egyptians built large boats to which they added square sails to harness the power of the wind. The Egyptian ships were made of wood and could hold a large cargo. At least twenty oarsmen were aboard to row the ship when sail power did not work.

The Phoenicians were the best of the ancient shipbuilders. Their merchant ships plied the seas carrying large cargoes. The Phoenicians made an important contribution to ships by adding two and three rows, or banks, of oars.

The Greeks, and later the Romans, expanded on the Phoenician idea by adding four and five banks of oars. Their ships were designed like modern ones, with curved

Egyptian ships were made of wood and were powered by sails, as well as at least twenty oarsmen. They also had a lot of room for cargo.

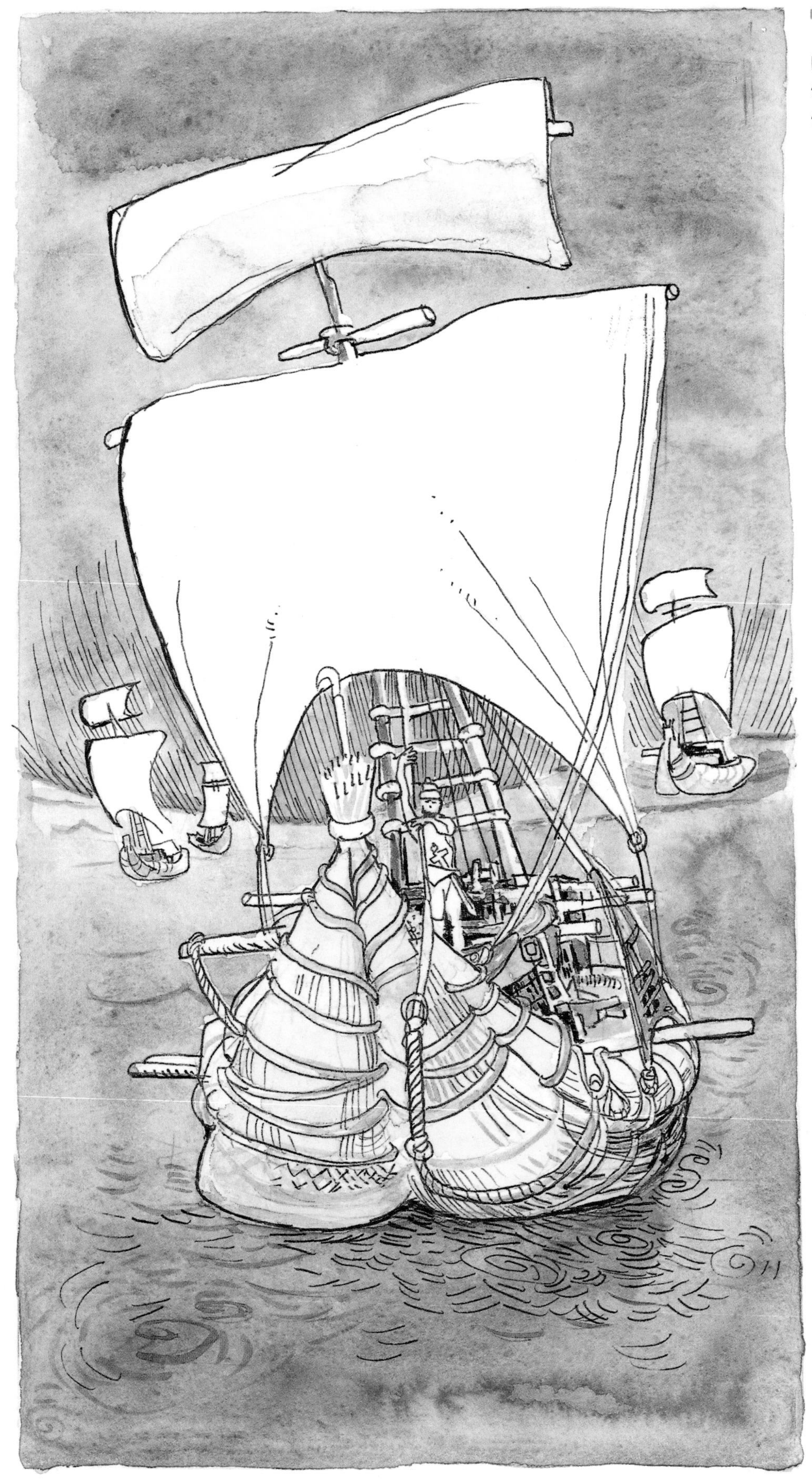

Phoenician ships were the best of the ancient ships. They had sails and two or three banks of oars.

timbers for the hull and a cabin at the back for the officers and captain. Viking ships were ocean-going vessels with both oars and sails.

During the Middle Ages, ships were similar to those of the Romans except that the rudder replaced the steering oars and the sides were higher above the water line. Sails were not used exclusively until late in the Middle Ages.

On August 17, 1807, forty years after James Watt invented the steam engine, the first commercial steamboat, the *Clermont*, was launched. Robert Fulton (1765–1815), an American engineer, designed the boat, and as a result, he is known as the "Father of Steam Navigation." The *Clermont* traveled 150 miles (240 km) up the Hudson River to Albany, New York, in thirty-two hours. Regular passenger

The Vikings sailed across the oceans in ships like these. Leif Erikson, a very famous Viking, is believed to have sailed to North America 400 years before Columbus! He named this continent Vinland. Archeologists believe that the Norse ruins at L'Anse-au-Meadow, in the province of Newfoundland, Canada, mark the place where Erikson originally landed.

service began a few days later. In more modern times, the development of motorships (diesel fuel ships) has made ocean travel much safer and faster than it was when steam engines were used.

Top: Paddle-wheel steamboats transported passengers up and down the great rivers of America, like the mighty Mississippi, throughout the 1800s. Bottom: The *Turtle,* a submarine built by David Bushnell, attacked a British warship during the American Revolution.

SUBMARINES

The submarine was first used during the time of the American Revolution, when David Bushnell's underwater vessel, named the *Turtle*, attacked the British warship, the *Eagle*, on September 6, 1776.

In 1801, Robert Fulton designed a submarine, the *Nautilus*, for the French government. The *Nautilus* could not travel very far, however, and Fulton designed an improved version in 1805 for the British.

During the American Civil War, the South had a submarine called the *Hunley*. The *Hunley* sank a Northern warship, the *Housatonic*, on February 17, 1864, in the harbor in Charleston, South Carolina. Unfortunately, the *Hunley* and its crew were lost during the conflict.

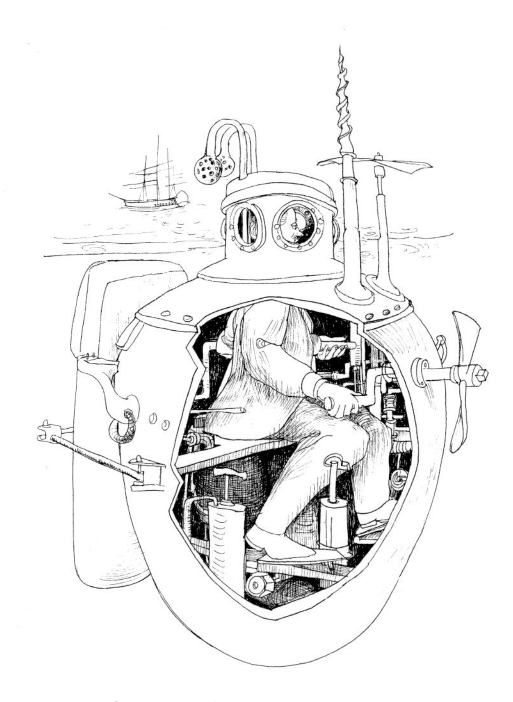

Below: This is a photograph of a German submarine, called a U-Boat.

When electricity became available, it became possible to build modern submarines. The first modern submarine was the United States Navy's *Holland*, designed by John B. Holland. The *Holland* sailed on April 11, 1900, but because it had to surface to recharge its batteries, it was not considered a true submarine.

The first true modern submarine was another U.S. Navy venture, also called the *Nautilus*. Put into service on September 30, 1954, the *Nautilus* used atomic fuel, and could stay underwater indefinitely.

The Locomotive

Before railroads, the only way to travel across vast stretches of land was by horse, using coaches or wagons. Horse-drawn carriages were often robbed or would break down, and wagon trains could be attacked or snowed in.

At right: In 1808, Richard Trevithick built this small locomotive, which he called "Catch Me Who Can." Located in central London, it traveled up to 10 miles (16 km) per hour on a circular track.

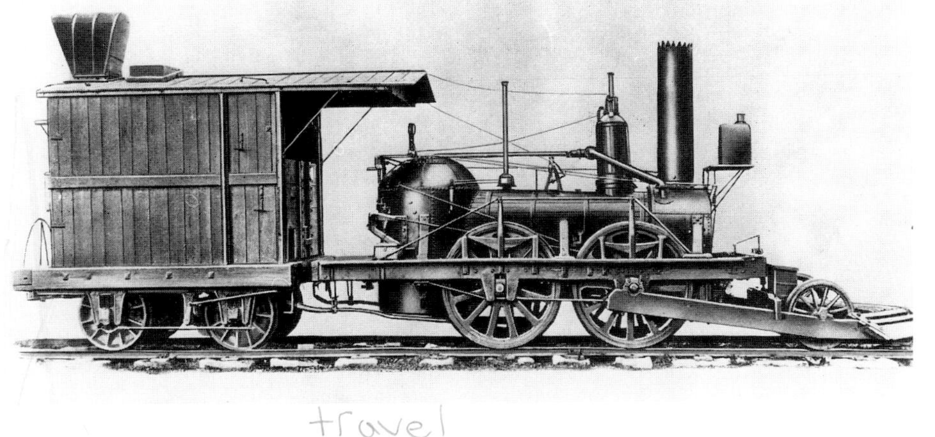

Smithsonian Institution

Robert Stevens brought this Stephenson locomotive to the United States from England in pieces. When he and a shipwright's apprentice named Isaac Dripps put the complicated locomotive together, they made several changes in the design. They named their new locomotive *John Bull*.

travel

The coming of the railroad meant people could move across distances faster, in larger numbers, and more safely.

To make rail travel possible, an engine with great power had to be invented. An Englishman named Richard Trevithick built a steam boiler in 1803 that moved on wheels. It moved so slowly, though, that he decided to put the boiler on rails, hoping to increase its speed. Three years later, he had a steam locomotive that could go about five miles (8 km) per hour.

Later, another Englishman, George Stephenson, built a locomotive to haul coal cars out of the mines. The locomotive began to operate on July 14, 1814.

In 1821, Stephenson and his son built a public railroad between Stockton and Darlington, England. On September 27, 1825, service began. The two then built another railroad between Manchester and Liverpool. Over the next hundred years, railroads spread rapidly across Europe.

As North America industrialized and settlers began to move west, railroads were built across the United States and Canada. An American, Robert Stevens, invented a new system of wooden ties, T-shaped rails, and hook-headed spikes. This rail system made travel by train much more comfortable.

Today, trains carry people around the countryside and are also used as a means of mass transit within cities.

Karl Benz built this four-wheel gas-engine automobile in 1887. He later joined with Gottlieb Daimler to build automobiles. His name appears on the Mercedes-Benz automobiles sold today.

THE AUTOMOBILE

Until this century, most of the world relied on animal-drawn carts or wagons to travel or transport things. Animals could not go very fast, and they had to be fed and cared for. Railroads made travel faster and more efficient, but railroads did not exist to every destination.

The history of attempts to develop powered land vehicles goes back several hundred years. Believe it or not, the plan for the first internal combustion engine (which powers an automobile) was developed by a Dutch scientist, Christian Huygens (Hoy-guns), in 1678. Unfortunately, he never built it. In 1680, Sir Isaac Newton built a toy propelled by the pressure of a jet of steam. In 1770, Frenchman Nicolas Cugnot (Koon-*yoh*) built a steam-powered machine, which was three-wheeled (like a tricycle), but only went two miles (3 km) per hour.

Inventor William Murdock (1754–1839), along with James Watt, who invented the steam engine (*see Chapter One*), built a steam-powered vehicle in Britain in 1781 and a steam-powered wagon in 1784. In 1786, British inventor William Symington (1763–1831) built a steam-powered carriage. Then in 1801, British inventor Richard Trevithick built the first steam carriage to carry passengers.

American inventor Oliver Evans received a patent on a steam carriage in 1789. In 1803, he built a self-propelled dredge, the first self-propelled vehicle to operate over American roads. By 1830, steam carriages regularly carried passengers over English roads, but accidents were very dangerous, with boiling water and glowing embers spewing from the engine. They were banned in 1831.

Such difficulties caused inventors to realize that steam was not the best way to power these machines. They began designing internal combustion engines. In 1860, a

Frenchman, Jean Joseph Lenoir (Zhanh Len-*wahr*), built an internal combustion engine. In 1863, he drove the machine he built six miles (10 km) in an hour-and-a-half.

In 1866, Eugen Langen (1833–1895) and Nikolaus Otto, a German, developed a more efficient gasoline-powered engine. Otto's chief engineer, Gottlieb Daimler, made an improved version of this engine, and built his first automobile in 1887. Karl Benz, another German, built an improved engine, and Benz and Daimler went into business together to build Daimler-Benz vehicles.

By this time, powered vehicles were called "horseless carriages." Eventually, they were called "automobiles," from the Greek word, *auto* (self), and the Latin word, *mobilis* (movable), because they were "self-movable."

The first automobiles were very expensive. In the United States, Henry Ford (1863–1947) decided to build automobiles anyone could afford. In 1908, he introduced the Model T, which was a standard until 1927. The automobile has revolutionized our lives, allowing us to travel great distances easily and comfortably.

The low-priced Model T, built by Henry Ford, made automobiles affordable for many people at the beginning of the twentieth century.

THE AIRPLANE

The concept of flight has fascinated humans for thousands of years, and we have long envied birds their freedom of movement above the ground.

While most people would credit the Wright Brothers with unlocking the secret of flight, these two men actually applied the concepts of inventors who lived long before them. In the 1500s, artist Leonardo da Vinci designed some "flying machines." He knew that to fly, one must imitate the gliding motion of birds, not the flapping motion. Unfortunately, he did not follow his own advice and spent many years trying to perfect a design that imitated the flapping motion of wings. Each version of his flying machine was heavier than the last, and he had no power source to lift it off the ground.

Da Vinci designed gliders, too, but he never attempted to build any of them. Modern tests have shown that these machines would have indeed flown. If da Vinci had pur-

These are some of Leonardo da Vinci's flying machine designs. Da Vinci wrote his notes backwards so other people could not easily read what he had written.

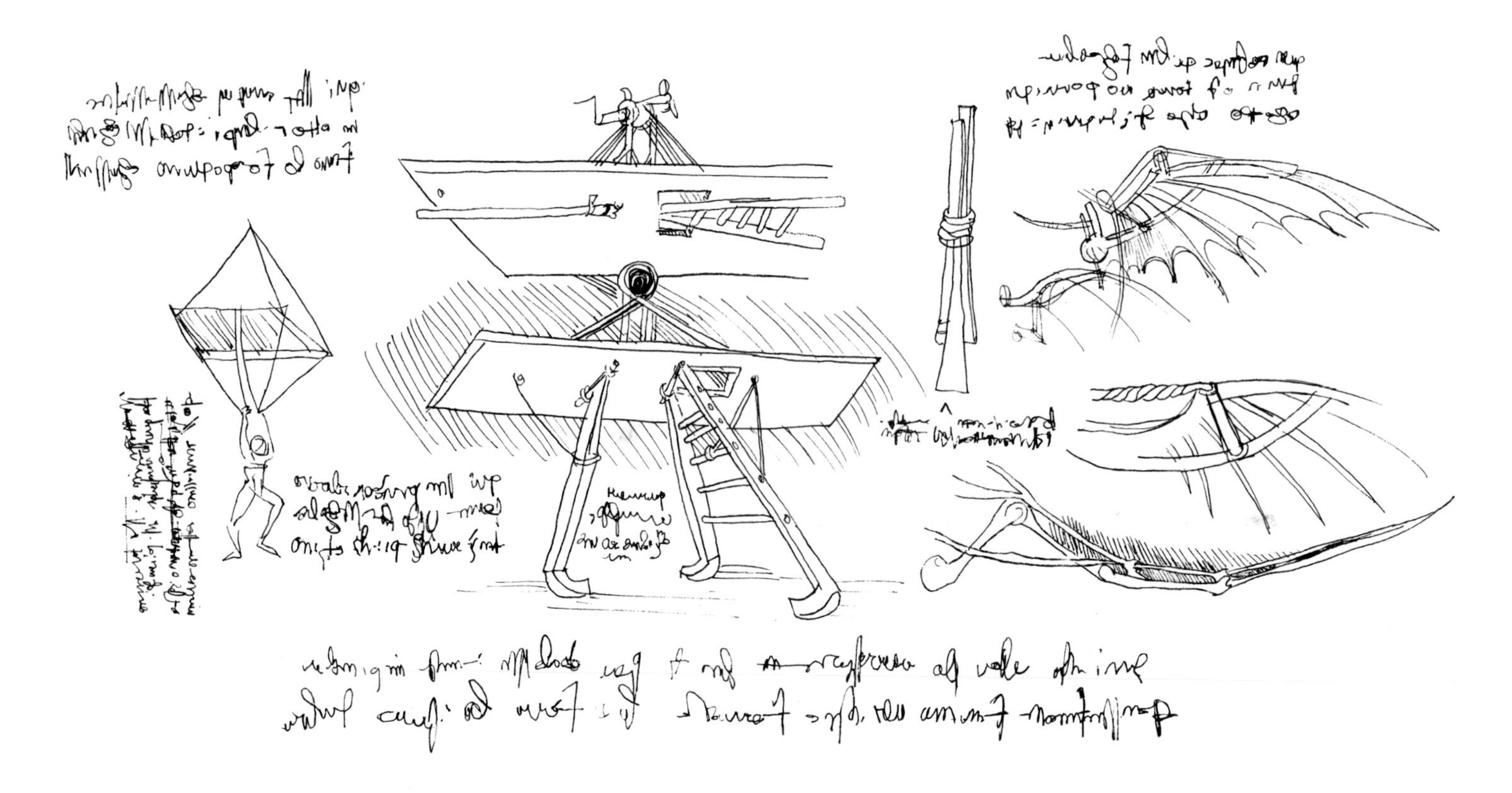

At left: Leonardo da Vinci invented the original concept for a helicopter in the 1500s. Not until the twentieth century, however, did anyone try to build such a flying machine.

sued his designs, he might have been the first human to fly. But despite his failure to build a successful flying machine, da Vinci is credited with being the first person to study flight in a scientific manner.

Da Vinci also developed the concept of the helicopter. Taking an idea from a Chinese toy, he designed an aerial screw that was supposed to whirl rapidly. Again, though, da Vinci's design was not adequate because it lacked a source of power.

Above: This modern industrial helicopter sits on the deck of an oil rig in the Gulf of Mexico.

A few hundred years later, in 1738, Swiss scientist Daniel Bournoulli (Ber-*noo*-ee) discovered the principle of the modern wing shape, or airfoil. During the 1800s, Briton Sir George Cayley developed the basic form of winged aircraft, and he flew the first glider in 1849.

Perhaps you have seen film footage of some of the early flying machines. Certainly not worthy of the name "airplane," these contraptions tried to imitate the flapping of a bird's wings, and all failed miserably in their attempts to get off the ground.

Orville and Wilbur Wright made a significant contribution to flight when they stopped trying to imitate the wing-flapping motion, and instead started to imitate the way eagles soar. They integrated the concepts of earlier inventors into their own ideas, and on December 17, 1903, at Kitty Hawk, North Carolina, their *Flyer* airplane made the first controlled and powered flight with a man, Orville, aboard. The flight only lasted twelve seconds, but they made three more flights that day, which lasted a total of thirty minutes.

Unfortunately, at first, no one believed them, but eventually, people realized the Wright Brothers were telling the truth. Even so, many considered the airplane a novelty, because no one could think of a practical use for it.

Then, in 1908, Orville Wright demonstrated a more powerful version of the *Flyer* to the American War Department. In 1909, the Wright Brothers got the first contract to build planes for the military. During World War I, airplanes proved themselves as useful weapons of war. But despite this success, no one gave serious thought to the commer-

The Wright Brothers' plane, the *Flyer,* made the first controlled and powered flight on December 17, 1903.

Courtesy Smithsonian Institution

Courtesy McDonnell Douglas

At left: Modern commercial jets, like this McDonnell Douglas MD-11, transport people all over the world every day.

cial use of airplanes until Charles Lindbergh flew solo nonstop from New York to Paris in 1927.

Still, many believed that airplanes would never be able to carry large numbers of people over long distances, because all that weight would require so much fuel that the plane could never get off the ground. Of course, these people had no way of knowing that the Germans would develop jet fuel in the 1930s, which would make commercial flight possible.

Now airplanes carry thousands of people all over the world every day. With airplanes, it is possible to have breakfast in Rome, lunch in Paris, and dinner in New York. Airplanes have indeed made the world a smaller place.

Below: The Mongols used rockets as weapons as early as A.D. 1232.

Rockets

When we think of rocketships, what comes to mind is science fiction and astronauts and modern space programs, but it was the Chinese who actually invented rockets, perhaps as long ago as 3000 B.C.

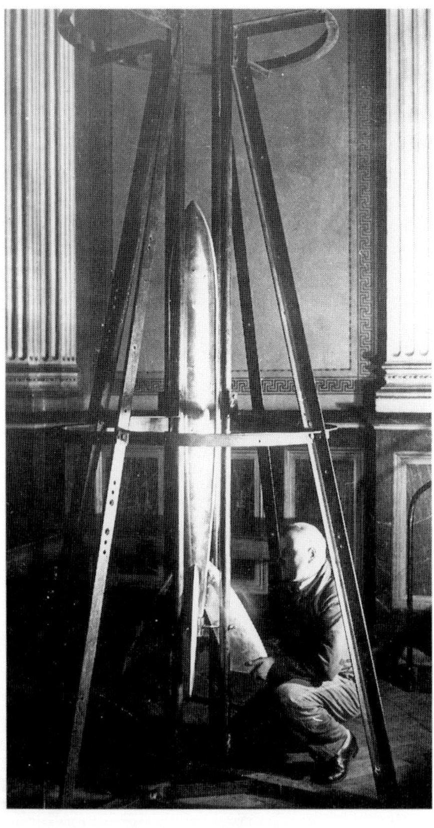
© Keystone/FPG International

Courtesy NASA

Top: This is a model of an Oberth moon rocket standing in its discharge tube. Bottom: On July 16, 1969, *Apollo 11* blasted off to begin humankind's first mission to land on the moon. The crewmen aboard included Neil A. Armstrong, Edwin E. "Buzz" Aldrin, Jr., and Michael Collins. Armstrong was the first to set foot on the moon, followed by Aldrin. Michael Collins remained behind in the command module.

Many scholars claim that the Chinese invented rockets around A.D. 1200, some 700 years earlier than the Western world's rocket development, and that the Mongols used them as weapons in A.D. 1232.

Rocketry as a science in the Western world began in 1895. At that time, two events took place. A Russian scientist, Konstantin Ziolkovsky (Tsiol-kof-skee), thought liquid fuels would be better than solid fuels to power rockets. A German scientist, Hermann Ganwindt, designed a very advanced spaceship. Neither of these men followed through with their ideas, however.

Then, in the United States, a man named Dr. Robert Goddard started to design rockets. He believed humans could go to the moon. On November 1, 1923, Goddard launched his first liquid fuel rocket, which soared 184 feet (55 m) into the air. On May 31, 1935, one of his rockets flew to 7,500 feet (2,250 m).

During World War II, the dictator of Nazi Germany, Adolf Hitler, decided that he wanted German scientists to pursue rocket technology as potential weaponry. Dr. Werner von Braun (1912–1977) headed Hitler's rocket program. He helped design the 46-foot-long (14 m) V-2 rocket, which was launched against Britain and France.

When the Germans surrendered in 1945, so did Dr. von Braun and his engineers, handing over the plans to the V-2. The United States brought von Braun and others to America. Because of these German scientists, America's rocket program made great advances.

Von Braun had always believed humans would go to the moon, and he wanted to build the rocket that got there. He convinced the U.S. government to give him that chance. He became the director of NASA in Houston, Texas. Von Braun was responsible for the Saturn 5 rocket that took the *Apollo 11* spacecraft to the moon.

The Space Shuttle

Sending a rocket into space was not the final goal of the space program. Scientists wanted a way for people to work in space. Helped by the knowledge gained from Werner von Braun and his team of engineers, American aerospace engineers were able to design and build the space shuttle.

Before the space shuttle, rockets were "throwaway" machines—that is, they were used only once. The shuttle, however, is the first reusable rocket system. Each shuttle will probably make 100 flights.

The real name of the space shuttle is the Space Transportation System (STS). Like a space-going truck, the shuttle can lift things into or return them from space.

The shuttle launches vertically, like a rocket, but lands like a glider; therefore, its design is much like that of an airplane. It is 123 feet (37 m) long and has a wingspan of 78 feet (23 m), about the size of a DC-9 commercial jet.

Besides carrying a standard crew of three astronauts, the shuttle can carry up to four mission specialists. While in orbit, the crew can perform a variety of experiments. Satellites can be launched or retrieved for repair and maintenance. Although the shuttle has no gravity, it has air, which means the astronauts do not need to wear spacesuits or helmets while inside.

A big problem in building the space shuttle was how to prevent it from burning up when it reentered the earth's atmosphere. Solving the problem led to other inventions: a new form of insulated, heat-reflecting tile and a certain kind of glue that holds the 32,000 tiles onto the shuttle.

The shuttle is a major milestone because it will carry the space station (*see Chapter 4*) into orbit several pieces at a time. Soon, not only will humans be able to work in space, but they will be able to live there, too.

Courtesy NASA

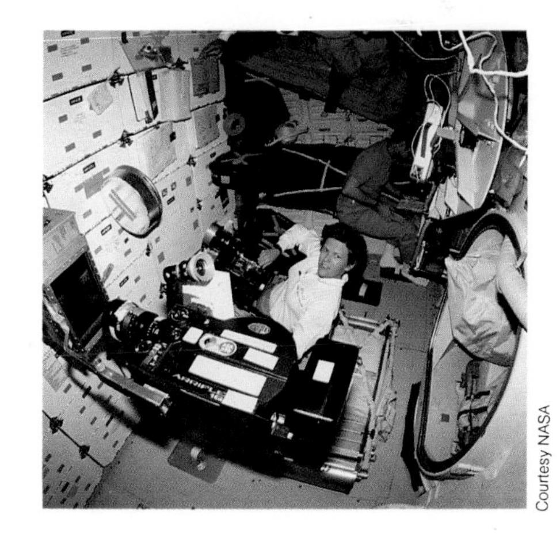

Courtesy NASA

Top: The space shuttle *Atlantis* soars into the sky. Although space shuttles launch vertically like rockets, they land horizontally like gliders. Bottom: On board the space shuttle *Discovery,* astronaut Kathryn Sullivan works on a camera, while astronaut Steven Hawley consults a checklist in the corner.

Breakthroughs in Communications

In prehistoric times, and even in some primitive cultures today, people communicated in only a few ways: face-to-face, with fire signals, or with drum messages. Communication was slow and limited only to those within sight or hearing distance. As time went on, several inventions came about that made communication easier.

WRITING

For centuries, there were no written languages and even spoken languages were still developing. By the time of the Neanderthals, humans told stories by drawing pictures on cave walls. They outlined their hunts and other events.

Writing as a method of recording the sounds of language made by the voice did not come about until shortly before 3000 B.C. This form of writing, created by the ancient Su-

Neanderthal man, the first *Homo sapiens,* drew pictures of hunting experiences on the walls of caves.

Although this particular painting was done by a modern painter, it is based on the art found in Europe in the caves of our early ancestors.

merians, who lived in Mesopotamia, contained symbols that represented whole words, or sometimes, whole phrases. These symbols are called *logograms*.

The Sumerian system of writing was part logogram and part *syllabic*, meaning the Sumerians also used symbols to represent syllables in words. We consider their "logosyllabic" system the first true system of writing.

But the Sumerian system was difficult to learn. Using a pure syllabic system is more efficient. Eventually, consonant and vowel sounds of words were given their own symbols. This is an alphabetic system of writing, such as we have in English. Although it may take many signs, or letters, to represent the sound of one word, the total number of symbols needed is very small. For instance, English has only twenty-six letters in its alphabet, and with these, we can write any word in the English language.

About 100 years after the Sumerians invented their writing system, the Egyptian system of writing, called hieroglyphics, was invented. As time went on, other peoples in the area developed their own writing systems. By 1500 B.C., semialphabetic writing was developed.

The Greeks borrowed their writing system from the Phoenicians. They took the important step of separating the consonants from the vowels, and by 800 B.C., they had arrived at a full alphabetic system.

It is sometimes difficult to understand how important the development of writing was to civilization. Before writing, people had to remember everything in their heads. Writing enabled people to record their thoughts and events for other people to enjoy and use.

Developing a writing system, however, was a first step. Writing was often accessible only to a privileged few. Not until the invention of the printing press in the 1500s was writing available on a mass scale.

THE PRINTING PRESS

Before the invention of the printing press, books and other documents were written and illustrated by hand. The Egyptians wrote on papyrus, a paper made from strips of reed pasted together, or on small tablets of wood or wax. The Romans made the first books by putting several sheets of parchment, a very thin leather, in a cover.

During the Middle Ages, few people knew how to read and write. Most books were written and illustrated by the monks in monasteries. Books were prized possessions.

Then, in the 1400s, Johann Gutenberg of Mainz, Germany, invented a printing press. Gutenberg's father was master of the Mint, where coins were made. Young Johann loved watching the goldsmiths stamp letters and figures on the coins, so he decided to become a metal worker.

Gutenberg's skill at metalwork led to his historic invention—the printing press—which featured metal let-

Johann Gutenberg invented a printing press in the fifteenth century. His printing press had movable type, metal letters that could be used over and over. In his printing shop, Gutenberg printed many copies of the Bible.

zu aaron. Streck deyn hannd vber dye fluß vnnd vber die bech vnd vber die brüch.vnd für

Although Gutenberg first printed a small book on his printing press, he became famous for printing the Bible. This Bible, called the Gutenberg Bible, was printed in Latin and had 1,282 pages. This is a page from a Gutenberg Bible.

ters that could be put in any order. Although he printed a small book first, he became famous in 1455 for printing the Bible.

Though history generally calls Gutenberg the first person to use movable type, the Chinese and Koreans actually invented printing around A.D. 400. The Chinese used blocks of wood with the letters and words carved out. They put ink on the letters and then stamped the page. Between A.D. 1041 and 1049, a man named Pi Sheng made the first movable type out of earthenware. Movable type meant that the letter blocks could be used over and over.

The invention of Gutenberg's printing press led in turn to the development of newspapers and printed books, which meant that for the first time, people could read about events instead of having someone tell them.

The sad part of Gutenberg's story was that he was poor, so he agreed to become partners with a lawyer named Johann Fust. Fust gave Gutenberg the money to build the press, but later demanded his money back. Fust knew Gutenberg would not be able to pay, so he took over the business. Gutenberg did not earn any money from his invention and died penniless twelve years later.

The Telegraph

Once writing was invented, people began to send messages over distances. Delivering messages and letters by runner, by rider on horseback, or by stagecoach took days, however, if not weeks or months.

In Massachusetts, a man named Samuel Morse, who was born in Charlestown in 1791, wondered how communications could be made faster. Since electricity could travel distances over a wire, Morse thought that electric messages could be transmitted over a wire, too.

Surprisingly, Samuel Morse was not a scientist, but an artist. Starting in 1832, he spent six years building a simple machine that used electricity to carry a message along a wire in a series of pulses. At the other end of the wire, a pencil, moved by an electromagnet, recorded the message. The message transmitted as sets of dots and dashes, each set spelling the letters of a word. This method of using dots and dashes is called Morse Code. The message it produces is called a telegram.

Very shortly after inventing the telegraph, Morse and his business partner, Arthur Vail, realized that the pencil was unnecessary, for anyone hearing the telegraph operate could tell the difference between the dots and dashes.

Samuel Morse poses next to his telegraph machine. Telegrams are sent worldwide to this day.

The United States Government was soon interested in Morse's new invention and constructed a telegraph line between Washington and Baltimore, which was forty miles (64 km) away. This line opened on May 14, 1844. Samuel Morse had the privilege of sending the first message. His first words were, "What hath God wrought?"

After this, the entire Western world wanted the telegraph. In 1850, an underwater cable was constructed between England and France. In 1866, another was placed under the Atlantic Ocean so telegraph messages could be sent between the United States and England.

The telegraph is still used, and one of the largest companies that sends telegrams today is Western Union.

This telephone from the late 1800s looks very different from the ones we use today.

THE TELEPHONE

The telephone is another method of instantaneous communication. American inventor Alexander Graham Bell (1847–1922) is credited with inventing the telephone, but he was not actually the first.

The telephone dates back to 1667, when English physicist Robert Hooke made a string telephone that carried sounds over a wire pulled tight. Around 1850, Sir Charles Wheatstone of England invented the acoustical phone. A musical box transmitted sounds from the cellar of a house to the second story of the same house using a wooden rod.

In 1854, French inventor Charles Bourseul (1829–1912) invented a telephone operated by electricity. In 1860, German physicist Johann Philip Reis (1834–1874) also invented an electric phone. The phone could not reproduce speech, however, and Reis gave up after a few tries.

Not until 1876 did Alexander Graham Bell, a man who taught deaf people how to talk, receive a patent for an electric phone. From his many experiments, he learned that

Alexander Graham Bell talks to a man in the next building on the telephone he invented.

only a steady electric current could transmit the human voice. The next year, he made the first phone that could transmit the human voice accurately. His phone consisted of a transmitter, a receiver, and a single connecting wire. He demonstrated it at the one-hundredth birthday exhibition of the United States in Philadelphia.

The telephone was an immediate hit. Over the next fifty years, nearly every household in the industrialized world had the new invention installed.

THE RADIO

In 1895, Guglielmo Marconi invented a radio. Radio soon became a worldwide method of communication.

The radio was invented in 1895 by an Italian, Guglielmo Marconi (1874–1937), winner of the Nobel prize for physics in 1909. Like most inventions, however, the radio was made possible by other inventions that came before it and by work that was done by others.

The first person to do work in the field of radio was English physicist James Maxwell, who published his theory on electromagnetic waves in 1873. However, Maxwell never pursued his theory.

Around 1888, a German, Heinrich (Hine-rik) Hertz, generated Maxwell's waves electrically. These waves were named "Hertzian waves" after him. Another scientist, David Hughes, invented a tube of loosely packed metal particles that conducted Hertzian waves.

In 1890, Marconi began to think about how he could invent a system in which telegrams could be sent without wires. In 1894, he read a newspaper article about electromagnetic waves and the Hughes tube. Marconi decided to try to improve Hughes' invention. He used the tube in some of his own inventions and made a crude radio.

In 1896, Marconi sent wireless signals about a mile (1.6 km). In 1897, he sent signals eighteen miles (30 km) away to a ship at sea. However, the Italian Ministry of Posts and Telegraphs was not excited about his new invention. Not giving up, Marconi traveled to England and showed his invention there. Unlike the Italian postal officials, the British officials liked Marconi's invention and told the British Government about him. The Navy was interested in Marconi's invention for ship-to-ship communication, so he demonstrated the wireless radio for them. Both the Navy and commercial shipping companies adopted the invention.

In 1899, Marconi started business use of the "wireless" (as the British still call the radio) between England and France. And in 1901, radio signals first crossed the Atlantic to North America.

All that Marconi transmitted, however, were the dots and dashes of Morse Code. On December 24, 1906, Christmas Eve, the first worded radio broadcast took place. The person responsible for the broadcast was not Marconi, but a Canadian inventor, Richard Fessenden. Once Fessenden made his historic broadcast, radio became a significant part of the communications world.

In 1920, in Britain, the Marconi Company broadcasted the first radio program. The company, along with five others, started making radios, or "wireless sets," and sold them as fast as they were made.

Also in 1920, the first commercial radio station in the United States began in Pittsburgh, Pennsylvania. In 1922, the British Broadcasting Company (BBC) was created to broadcast programs in Great Britain. The BBC still operates that radio station and two television stations as well.

This picture is a close-up view of Marconi's radio. Made up of many metal parts and tubes, it bears no resemblance to radios today.

Paul Gottlieb Nipkow invented the first television in 1884. Almost 60 years passed before the first commercial television station began operating.

TELEVISION

Television has had a tremendous impact on the world. With television, people can see events taking place as they occur all over the world. In 1991, the Persian Gulf War pitting the United States and its allies against Iraq unfolded before our eyes as if we were actually there ourselves.

Before television, people heard the news on the radio or read about the news in the papers hours or days after it happened. Not so long ago, people went to the movie theater to see the news on "newsreels."

This early television looks very different from the televisions we have today.

The first television, which was really a way of scanning images, was invented by German inventor Paul Gottlieb Nipkow in 1884. This flat, circular disk scanned an image from the top to the bottom and transmitted it electrically. The design, however, had problems.

In England, in 1925, John Logie Baird invented the first device that transmitted television signals. Other successful television broadcasts were made in 1927 in England and in 1930 in the United States.

In 1933, a Russian immigrant to America named Vladimir Zworykin invented the iconoscope, which picked up changes in light intensity and turned these into electric charges. In 1934, Philo Farnsworth invented the television camera, making commercial television possible.

On April 30, 1939, during the New York World's Fair, scheduled broadcasting began. World War II interrupted the broadcasting, but it started again in 1946 when the war was over. By the end of that year, twelve commercial stations were operating in the United States. Today, nearly every home in every town has a television.

SATELLITES

If you stare up at the sky on a clear summer night, you will eventually see a satellite fly over as it orbits the earth.

There are two main kinds of nonmilitary satellites in space: communications satellites and weather satellites. Neither type would be possible without the research done by the scientist Werner von Braun and those who followed him. Rockets, which were invented by von Braun, are how we get satellites into space.

Communications satellites allow us to send and receive live television programs and telephone calls all over the earth. They also keep ships in contact with each other and with their shore-based offices.

Because of communications satellites, millions of people can enjoy events like the Olympics as they happen, no matter where in the world they are located.

Communications satellites work by receiving microwave signals from an earth station, amplifying (increasing) them, and then sending them back to the earth. Any station in the area where the signals are beamed can pick them up. There are earth stations in more than 100 countries.

The Soviet's *Sputnik* was the first satellite, launched in October 1957. The United States did not launch its first satellite until 1958. Other satellites were launched in the 1960s, which were the first with "synchronous" orbits (they move at the same speed as the earth turns).

Courtesy NASA

Courtesy NASA

Top: The Syncom IV-5 communications satellite is launched from the cargo bay of the space shuttle *Columbia*. Bottom: The same communications satellite orbits over Zaire in this picture taken by crew members aboard the space shuttle *Columbia*.

Weather satellites transmit information about weather patterns all over the earth. The Geostationary Operational Environment Satellites (GOES), built by Hughes Aircraft Company, send infrared pictures of the earth every night to weather stations all over the world.

There are several GOES satellites, each monitoring a different part of the earth. GOES send pictures back to the earth every thirty minutes. These are the pictures we see on the television weather reports and in the newspaper.

Receiving pictures every thirty minutes lets meteorologists (weather scientists) trace storms and provide warnings for events like floods and hurricanes. GOES can monitor the eruptions of volcanoes and track their plumes of volcanic ash. Weather satellites also help scientists observe changes in the oceans, study the effects of drought, find mineral deposits, measure destruction of the rain forests, study nuclear test sites, and pinpoint earthquake faults.

Weather satellites have given us the opportunity to have some "control" over our weather, and they have often helped us to predict many natural disasters.

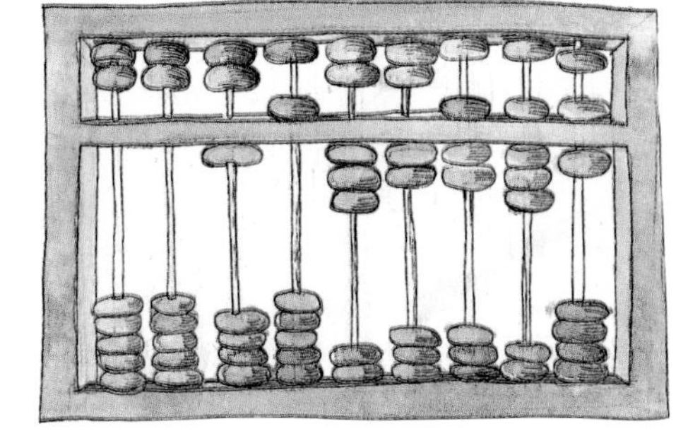

The abacus was the world's first computer. The Chinese invented it in 600 B.C. The abacus is still used in many Asian countries today.

THE COMPUTER

Today, many of us cannot imagine a world without computers. They perform tasks in seconds that would have taken days by hand.

And yet, the computer is not as recent an idea as one might think. The Chinese invented a form of a computer in 600 B.C. Their computer was called an abacus, and with it a person could perform mathematical calculations very quickly. The abacus is a small wooden frame with straight pegs or wires across it. Calculations are made by moving the beads across the pegs. This "computer" was in use for 2,500 years, and it still is in some areas of the world.

Then, in 1643, a French mathematician named Blaise Pascal invented the adding machine. Around 1673, Gottfried von Leiber, a German, invented a calculator that could add, subtract, multiply, and divide.

Two hundred years later, Charles Babbage came up with an innovative idea for a machine he called an "analytical engine," because its purpose was to analyze mathematical problems. Unfortunately, Babbage could not build his machine because two parts that he needed had not been invented yet—the vacuum tube and the cathode ray tube (the picture tube in a television set or computer monitor).

In 1946, the University of Pennsylvania made the world's first true computer. Built for the United States Government, this computer had 20,000 vacuum tubes and took up a large portion of a building.

As the space program progressed, the need for smaller electronic parts grew, which, in turn, reduced the size of computers. At the same time, their power increased. The real turning point, however, was the invention of the transistor, which made vacuum tubes unnecessary.

Computers continued to get smaller and more powerful. Until the late 1970s, computers were so expensive that only corporations could afford them. Then, in 1983, IBM Corporation introduced the Personal Computer, or PC, a small computer. The PC revolutionized our society, because it made computers affordable for individuals.

Computers perform tasks many times faster and more accurately than can often be done by hand. The ability to do complex accounting, repetitive typing chores, typesetting of brochures and other advertising materials, in addition to storing and retrieving large amounts of information, has made the PC a permanent fixture in almost every office, in many homes, and even in some automobiles and many electronic products.

© Keystone/FPG International

© FPG International

Top: This picture from 1935 shows Mr. A. Porter working on his model of the calculating machine he built. Calculations that would take mathematicians several days to complete took this machine only minutes to complete. Bottom: Unlike the first computers, which were extremely large, today's computers are small enough to fit on the top of a desk.

Breakthroughs in Living Modes

Millions of years ago, our ancestors lived in trees. During the Stone Age, they lived in caves and hunted for food. After a long period of time, some groups of people began to live in one place and either grow their own food or trade for it. Eventually, the places those people lived in grew into cities. Someday, humans will be living in cities located in space. Human progression from the caves to cities in space is the story of invention.

AGRICULTURE

Before the invention of agriculture, or farming, people moved around, hunting for food or gathering plants. The men hunted wild animals, while the women and children gathered berries, roots, and other plants. The food the women gathered made up 70 percent of their diet. Because people were dependent on plants, they limited their groups, or bands, to no more than twenty-five people.

Eventually, people learned to take care of plants as they grew, then, they started to grow plants on their own. They developed the hoe and digging stick as tools and began to raise animals to eat.

Catal Huyuk, the first known farming community, was built about 7,700 to 8,500 years ago. People lived in mud-brick houses that were attached to each other.

During the Neolithic Era, the development of agriculture allowed people to settle in one area and build a permanent settlement—a village. The first known farming community, Catal Huyuk in Turkey, began about 7,700 to 8,500 years ago. The people in this town lived in mud-brick houses that were attached to each other. They painted the walls with scenes of hunters, dancers, and animals.

Irrigation, developed around 4000 B.C., diverted natural sources of water to dry areas of land, allowing cultivation to develop in new places. By 3000 B.C., the Mesopotamians and Egyptians had developed the wooden plow, so they could farm larger areas of land. Other early agricultural societies have been found in India. Agriculture began in many places at the same time and eventually spread around the world from Turkey, Egypt, and Mesopotamia.

Agriculture was only adopted by cultures whose standard of living it improved. Some cultures tried farming, found that their living standard went down, and went back to hunting and gathering. Even though there may have been plenty of food to hunt and gather, their area may not have been suitable for agriculture, so they rejected it.

Where farming communities developed, populations were able to increase due to new methods of irrigation and food control. Over time, agriculture brought about many other important changes, such as different occupations, political and religious organizations, road and building construction, laws, markets, and eventually, cities.

CITIES

Farming communities were one way cities evolved, but cities also grew at good locations along trade routes. For instance, one city developed next to a mountain rich in obsidian, a type of volcanic glass. The people traded the obsidian, used for razor-sharp tools, for food. This new pattern of living in one place caused the old social organization of wandering tribes to break down.

All of the first cities grew before the invention of writing. The oldest city, Tell Mureybit, in the Middle East, is 8,000 years old, which means it came about during the Bronze Age. During the Bronze Age, humans developed the art of making things from metal, called metallurgy.

Tell Mureybit had seventeen levels of houses, and research since World War II has discovered that this city's inhabitants were not farmers. They ate wild grain and meat from animals they hunted. Other cities similar to Tell Mureybit developed in Egypt, China, India, and later in Europe, Africa, Mexico, and Peru.

Cities are important for several reasons. First, like farming communities, cities allowed people to use their time for things other than food gathering. Some people continued to grow the food, while others made the clothes, and still others made the cooking utensils.

Second, cities provided safety. While a band of twenty-five people might be vulnerable to food shortages or attack by an enemy or wild animals, cities with thousands of people and with walls around them were not as vulnerable.

Some scholars now think that agriculture was the result of cities, not the other way around. Cities do not require agriculture, but only the presence of a large amount of food. Perhaps the rise in population began with the receding of the Ice Age. Lush grass grew, and many wild animals roamed the areas. Perhaps the population increase caused the people to build a town in a pleasant, livable place.

A source of food in one place also allowed for survival of the weak. The young, old, and sick, who would have died on the search for food, were able to survive. People lived longer, had more children, and the villages grew.

Tell Mureybit, the world's oldest city, was built 8,000 years ago. It had seventeen levels of houses.

When the best areas were taken, people settled in less favorable spots. Since food was harder to get in these areas, people provided special services, like irrigation, defense, and business, in exchange for food. Once cities got big enough, government came about. A ruling class grew. Cities forced the development of legal systems, training of defense troops, and grain storage.

What really makes a city a city is the number of people who live there. An urban center is created when more people in the area live there than on the farms. By 2500 B.C., many large urban centers existed.

SPACE STATION

For years, science fiction writers and fans have dreamed of the day when humans could leave the earth to live in outer space. That dream came true for the Russians in 1972, with the launch of their first space station. Ever since then, cosmonauts (Russian astronauts) have been living in space, with each crew staying six months.

Humans may one day live on the moon or on other planets in colonies like these. They are enclosed to provide an artificial environment with breathable air.

The Americans attempted their first space station with *Skylab* in 1973. Unfortunately, *Skylab* could not stay in orbit and fell to earth about five years after it was launched. Hope for a permanent, manned presence in space will be realized before the year 2000, with the launch of the space station *Freedom.* This space station will have people permanently living and working in space.

More than twenty shuttle flights will be required to put the space station together in space. It is the biggest international construction project ever attempted. Currently, four space agencies are involved in this undertaking, with each contributing a portion of the station. The United States is building the habitation module, in which the astronauts will live; a laboratory module; and several other elements. The European Space Agency is constructing a laboratory module, too, and what is called a Man-Tended Free Flyer (MTFF), which will orbit around the earth with the station. Canada is building a device called a Mobile Servicing System (MSS), which is a type of robotic arm that will be used in constructing and operating the space station. Japan is building the Japanese Experimental Module (JEM) and an attached "logistics" module. Although each agency owns its own modules, they will be used by all.

In its first phase, the space station will be an orbiting laboratory. The astronauts will be able to observe the earth and perform experiments in life sciences, physics, astronomy, new manufacturing techniques, and materials processing. Some of the things that may be developed are pharmaceuticals, or medicines, for treating diseases like AIDS. Some pharmaceuticals cannot be developed on earth due to our gravity, which makes the medicines separate. In space, however, the elements can blend perfectly. Many other inventions, some that may change the way we live, will be possible with the space station.

Courtesy NASA

Space station *Freedom* is humankind's first real space station and will be ready for use in 1999. *Freedom* is small, but it marks our first steps toward a space community. Eventually, people will be born, live, and die in outer space.

BIBLIOGRAPHY

Aaseng, Nathan. *The Inventors: Nobel Prizes in Chemistry, Physics, and Medicine.* Minneapolis, Minnesota: Lerner Publications Company, 1988.

Adler, Irving. *Fire in Your Life.* New York: The John Day Company, 1955.

Barrett, N.S. *Robots.* New York: Franklin Watts, 1985.

Cook, David C. *Inventions That Made History.* Toronto: Longman's Canada Limited, 1968.

Cooper, Margaret. *The Inventions of Leonardo da Vinci.* New York: The MacMillan Company, 1965.

Dalby, A. Royce. "Let the Construction Begin." *Ad Astra,* Volume 2, Number 7, (July, August) 1990: 15.

Design News. "Lasers Stake Out New Frontiers." *Design News.* (March 13, 1989): 22–33.

D'Ignazio, Fred. *Working Robots.* New York: Elsevier/Nelson Books, 1982.

Dineen, Jacqueline. *Twenty Inventors.* New York: Marshall Cavendish, 1988.

Douglas Aircraft Co. *Personal Interview with Elayne Bendel.* Long Beach California, 1990.

Encyclopedia Brittanica. "Satellite Communications." *Telecommunications System* (1989): 504–505.

Federal Aviation Administration. *Pilot's Handbook of Aeronautical Knowledge.* ASA Publications, 1980.

Flying Magazine. *America's Flying Book.* New York: Charles Scribner's Sons, 1972.

Frazer, Lance. "Sensory Reparation." *Ad Astra,* Volume 2, Number 7, (1990): 27.

Galton, Lawrence. "A New Age of Medical Miracles." *Consumers' Digest* (March, 1986): 57–60.

Good Housekeeping. "Miracle Techniques to Replace Major Surgery." (June, 1987): 212–213.

Graham, Ian. *Inventions.* New York: The Bookwright Press, 1987.

Greene, Carol. *Robots.* Chicago: Childrens Press, 1983.

Hamblin, Dora Jane et al. *The First Cities.* New York: Time-Life Books, 1973.

Hawke, David Freeman. *Nuts and Bolts of the Past: A History of American Technology, 1776–1860.* New York: Harper & Row, 1988.

Hooper, Meredith. *Everyday Inventions.* New York: Taplinger Publishing Co., Inc., 1976.

Howell, F. Clark. *Early Man.* New York: Time, Inc., 1970.

Jeuneman, Frederic B. "City Above the Clouds." *Research & Development,* Volume 26, (August, 1984): 19.

Knight, David C. *Robotics: Past, Present, and Future.* New York: William Morrow & Company, 1983.

Lavrakas, Paul. "Laser: The Healing Light of Medicine." *Consumer's Research,* (October, 1985): 11–15.

Lenga, Rosalind. *The Amazing Fact Book of Planes.* A & P Books, 1980.

Marsa, Linda. "No Knife Surgery." *Redbook.* (September, 1988): 28.

Morse, Joseph Laffan, ed. *Funk and Wagnalls New Encyclopedia.* Volumes 1, 3, 10, 12, 20, 21, 22, 23, 25. New York: Funk & Wagnalls, 1973.

Noreen, Gary K. "Dawn of a New Era in Mobile Satellite Communication." *Telecommunications,* Volume 20 (June, 1986): 61–63.

Otto, Dixon P. "Observing the Oceans from Orbit." *Space World* (April, 1983): 22–27.

Parker, Edwin G. "Future Perspectives on Satellite Communication." *Telecommunications,* Volume 21, (August, 1987): 47–48.

Presence, Peter, ed. *Encyclopedia of Inventions.* Secaucus, New Jersey: Chartwell Books, Inc., 1976.

Sharma, Ranjana. "Satellite Network Design Parameters and Trade-Off Analysis." *Telecommunications,* Volume 21, (June, 1987): 36–38.

Silverstein, Dr. Alvin, and Virginia B. Silverstein. *The Robots are Here.* Englewood Cliffs, New Jersey: Prentice-Hall, Inc. 1983.

Sonenclar, Robert. "Lasers on Target." *Financial World,* (23 January/5 February 1985): 20–23.

INDEX